TEST YOURSELF

STATISTICS

by

A. Stephen

CELTIC REVISION AIDS

CELTIC REVISION AIDS

Lincoln Way, Windmill Road,
Sunbury on Thames, Middlesex

© C.E.S. Ltd.

First published 1980

ISBN 017 751242 3

Printed in Hong Kong

FOREWORD

This book of self-test questions is a useful aid to a quick and thorough revision covering all the topics on the G.C.E. O Level and various C.S.E. syllabuses in Statistics. Most of the examinations in O Level and C.S.E. Statistics now have a number of questions requiring short answers and the majority of the questions in this book are taken from past examination papers.

My thanks are due to the A.E.B. and to the East Anglia Examinations Board for permission to reproduce questions from their past papers although those Boards, of course, accept no responsibility for the answers.

1. Outline the basic rules to be followed in tabulation.

2. The following table shows the investments, in millions of pounds, in certain assets by the Post Office Savings Bank during the 3rd quarter of 1967. Fill in the blank spaces (a)–(f).

	Investments made	Investments realised	Net investment
British Government securities	9.63	4.05	5.58
Local authorities: Quoted securities	7.08	(a)	(b)
Mortgages	(c)	0.55	0.05
Temporary loans	(d)	1.13	(e)
Total	17.96	(f)	11.50

(Source—Financial Statistics.)

3. State <u>six</u> general rules to bear in mind when designing a questionnaire which members of the public will be asked to fill in.

4. State one reason why the question "Have you any children?" is unsuitable for use in a census form addressed to the head of every household.

1. (a) Tables should be headed and the heading should indicate the sort of information provided. Time and place should be stated.
   (b) Units (thousands, kilowatt, hours, etc) should be stated.
   (c) Any complications should be added in footnotes.
   (d) Source of the information (e.g. HM Treasury) should be stated.
   (e) Rule off columns and lines and leave spaces to aid clarity.
   (f) Rates, ratios and percentages should not be used without giving the information on which they are based.

2. (a) 0.73
   (b) 6.35
   (c) 0.60
   (d) 0.65
   (e) −0.48
   (f) 6.46

3. (a) The number of questions should be kept to a minimum.
   (b) Questions should be simple and unambiguous.
   (c) Factual answers should be encouraged.
   (d) Emotive words should be avoided.
   (e) "Loaded" questions should be avoided.
   (f) Do not take name and address unless really necessary.

4. The answer "yes" reveals insufficient information. "How many children have you?" or "Give number, sex and age of all children" would be better.

5. State briefly two reasons why a firm may undertake a pilot enquiry before proceeding with a large-scale survey.

6. Distinguish between primary and secondary data.

7. Explain briefly why non-response causes difficulty in a sample survey.

8. State one advantage and one disadvantage of postal surveys.

9. A questionnaire, directed at housewives, was to be issued by a manufacturer of tinned foods and the following items were suggested:
   (a) What is your age?
   (b) Have you any children?
   (c) How much does your husband earn per week?
   (d) Make a list of the tinned foods in your house.
   Comment briefly on the suitability, or otherwise, of each item and make suggestions as to how you think these items could be improved.

10. Explain briefly what is meant by a biased sample, and give an example where it might arise.

11. Define a random sample.

5. (a) Misunderstandings can be eradicated in the full survey.
   (b) It indicates whether you are going to obtain the information you need.

6. Primary data is obtained by direct observation e.g. measurement or counting; secondary data comes from statistics already in existence.

7. Non-response makes the sample less valid and smaller and it may be that those not responding have a common characteristic.

8. Advantage: reaches everybody; Disadvantage: many forms not returned.

9. (a) Unsuitable (too vague). Suggest "Give your date of birth".
   (b) Unsuitable (too vague). Suggest "Give number, sex and date of birth of all offspring".
   (c) Unsuitable (too vague). Suggest "State your husband's basic weekly wage".
   (d) Unsuitable (too vague). Suggest several easier questions e.g. "How many tins of food did you buy last week?".

10. A biased sample is one which is not properly representative of the whole group. It could arise by selecting names from a telephone directory since such a method excludes the poorer sections of the community.

11. A random sample is one in which at each stage all members of the population, not already selected, stand the same chance of being included.

12. Imagine you are in a large mixed Comprehensive School, and are asked to find out which television programmes are most popular with the pupils. You decide to ask a limited number of pupils to fill in questionnaires, using a method of stratified sampling to select the pupils. State very briefly how you would obtain your samples.

13. What is "multi-stage" sampling?

14. In each of the following cases state whether the sample taken is a random sample:
    (a) Farms are selected from a particular area by repeatedly sticking a pin in a large-scale map.
    (b) Schoolchildren are selected from a particular school by taking every 10th name on an alphabetical list.

15. "Every member of the parent population must have the same chance of selection." Name the statistical process to which this statement refers.

16. Give an example of a discrete variable.

17. Give an example of a quantitative variable.

18. What is meant by a continuous variable?

19. What is a qualitative variable?

20. What distinguishes a discrete variable from a continuous variable?

12. Stratified sampling involves grouping the sample into sub-groups or strata. Pupils could be divided into male and female and each of these groups could be further divided into age groups and each age group could be further divided into house or form or class groups and a sample taken from each sub-group.

13. Multi-stage sampling involves taking a sample from a large group and then taking a second stage sample of small groups (e.g. counties and then towns).

14. (a) Yes.
    (b) No, since after each selection the next 9 are excluded from a chance of selection.

15. Random sampling.

16. Number of children in a family.

17. Salaries of teachers.

18. One which is not capable of exact measurement e.g. temperature.

19. One which is not capable of numerical expression.

20. A discrete or discontinuous variable is one where each successive value can be exactly measured and where there are distinct breaks between each one whereas a continuous variable is one where there are no real breaks between one value and the next and where each successive value cannot be exactly measured.

21. Name a characteristic of <u>children</u> that is
    (a) A qualitative variable
    (b) A quantitative variable

22. During a survey vehicles are classified into the following categories: A. Motor-cars; B. Lorries and vans; C. Buses and coaches; D. Motor-cycles.
    (a) Is the basis of this classification qualitative or quantitative?
    (b) Is the category of a vehicle a discrete or a continuous variable?

23. The variable x represents the total number of people using a municipal swimming bath in a day. Classify x as
    (a) Continuous or discrete
    (b) Qualitative or quantitative

24. Define (a) class interval,
           (b) class limits,
           (c) class boundaries.

25. The masses of fruit of a certain species were determined to the nearest gram and tabulated as follows:

Mass (g)	10–12	13–15	16–18	19–21
No. of fruit	8	13	16	6

What is the class interval?

26. Observations of a continuous variable, which can take positive values only, are classified into four classes whose central values are $\frac{1}{2}$, $1\frac{1}{2}$, 3 and 7. What are the sizes of the corresponding class intervals?

21. (a) Sex
    (b) Number of grandparents

22. (a) qualitative
    (b) discrete

23. (a) discrete
    (b) quantitative

24. (a) The range of values used in defining a class (e.g. £10–£60).
    (b) The upper and lower limits of each class interval.
    (c) The "true" class limits (the class boundary is the average of an upper class limit and the subsequent lower class limit).
    In the table:

$$0-50$$
$$51-100$$
$$101-150$$

    the class interval is 50; the class limits are (in the second line) 51 and 100; the class boundaries (in the second line) are $50\frac{1}{2}$ and $100\frac{1}{2}$.

25. 3.

26. 1, 1, 2, 6

27. The heights of a group of men were measured to the nearest cm and recorded as follows:

Height (cm)	No. of men
155–159	3
160–164	6
165–169	9
etc	etc

State (a) the limits between which the actual heights of the six men in the class 160–164 must lie, (b) the class interval.

28. What is the lower limit to the following calculation if the numbers are given in significant figures?

$$\frac{3.8 \times 2.9}{1.2}$$

29. Some wooden rods were measured to the nearest centimetre, and the results were tabulated as follows:

Length (cm)	Frequency
12–14	18
15–17	32
18–20	16

What is the length of the shortest rod that could be included in the class "15–17 cm"?

30. What is meant by "error" in statistics?

31. Explain the difference between "absolute" and "relative" error.

27. (a) $159\frac{1}{2}-164\frac{1}{2}$
     (b) 5

28. 8.6 $\left(\text{i.e. } \dfrac{3.75 \times 2.85}{1.24} \right)$

29. 14.5cm

30. The difference between the actual figures being considered and their true value.

31. Absolute error is the difference between the true value of a figure and the rounded figure employed; relative error compares the magnitude of the error and the actual figure (usually expressing the error as a percentage of the actual figure).

32. Each of two lengths is measured correct to the nearest centimetre; their sum is 19cm. Each of a further eight lengths is measured correct to the nearest 5 millimetres; their sum is 83cm. What is the maximum possible error in the arithmetic mean length of 10.2cm?

33. What is the upper limit to the following calculation if the numbers are given in significant figures?

$$\frac{1.6 \times 3.7}{0.8}$$

34. A motorist notes that, according to the clock and the distance-measuring instrument in his car, he has taken 1 hour to drive 50km. He concludes that his average speed was 50km/h. If each of his instruments is accurate to 1% only, between what limits does his true average speed lie?

35. In a certain town the birth rate, to three significant figures, was 12.7 per thousand when the population, to two significant figures, was 25,000. Calculate extreme values for the number of births.

36. Draw a sketch of a cumulative frequency curve.

32. 2.1cm (Sum of first two could be 18 or 20; sum of further eight could be 63 or 103; sum of all ten could be 81 or 123; mean of all ten could be 8.1 or 12.3)

33. 8.2 $\left(\text{i.e. } \dfrac{1.64 \times 3.74}{0.75} \right)$

34. 49km/h and 51km/h since time could be 1.01 hours or 0.99 hours and distance could be 50.5km or 49.5km and therefore

minimum speed $= \dfrac{49.5}{1.01} = 49$km/h and

maximum speed $= \dfrac{50.5}{0.99} = 51$km/h.

35. 309.925 and 324.857 since birth rate could be 12.65 or 12.74 and population could be 24,500 or 24,499 and therefore

minimum births $= \dfrac{12.65 \times 24,500}{1,000} = 309.925$;

maximum births $= \dfrac{12.74 \times 25,499}{1,000} = 324.857$.

36.

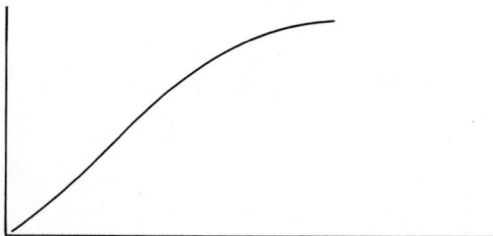

37. *Draw a sketch representing a normal distribution.*

38. *Sketch a curve showing positive skew.*

39. *Sketch a curve showing negative skew.*

37.

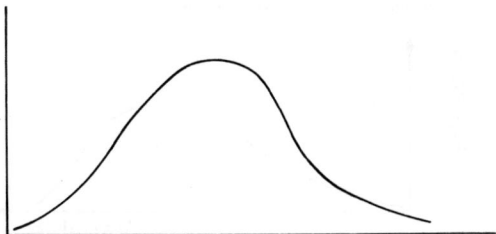

38.

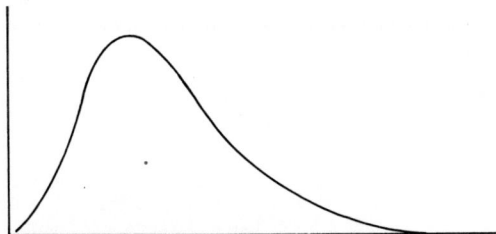

39.

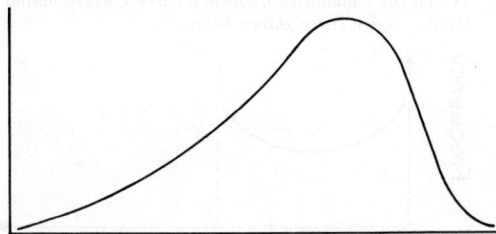

40. Sketch the cumulative frequency curve corresponding to the frequency distribution shown below:

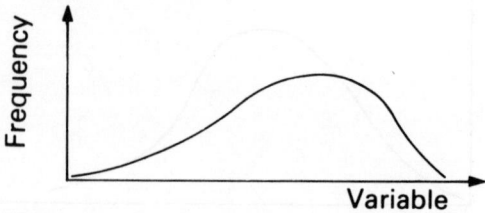

41. Sketch a frequency distribution for which the variable is much more likely to take extreme values than central values.

42. Sketch the cumulative frequency curve corresponding to the frequency curve shown below:

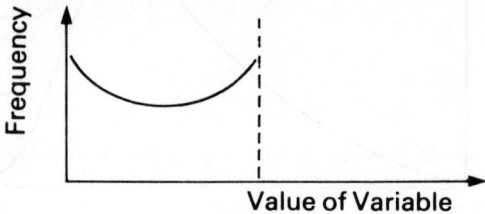

40.

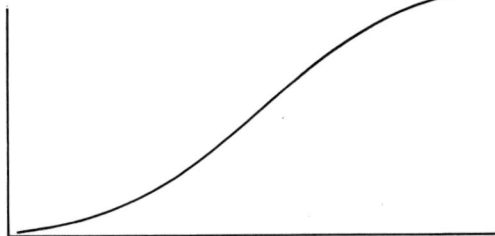

41.

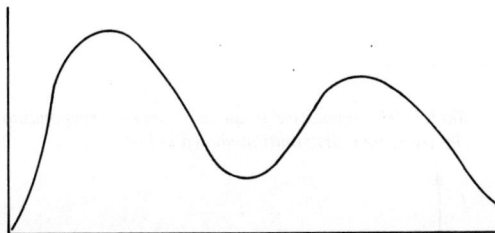

42.

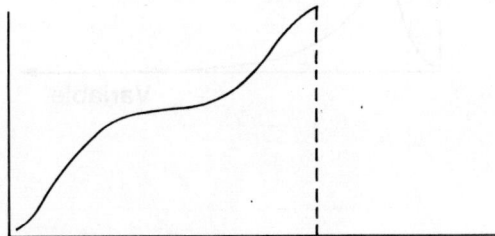

43. *Sketch a frequency distribution in which the mode is larger than the mean.*

44. *Sketch the cumulative frequency curve corresponding to the frequency distribution shown below:*

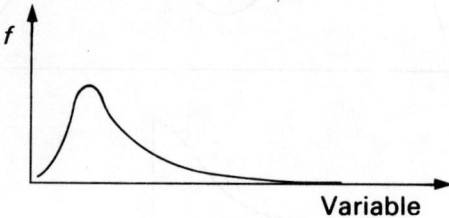

f

Variable

43.

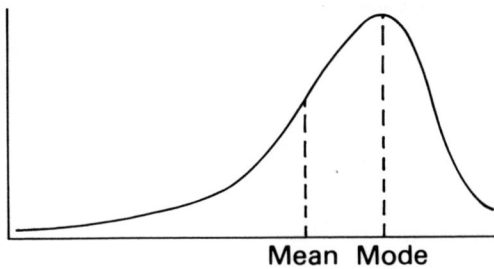

Mean Mode

44.

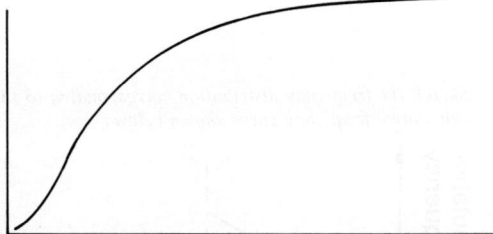

45. *Sketch frequency distributions which are (a) J-shaped, (b) bi-modal.*

46. *Sketch the frequency distribution corresponding to the cumulative frequency curve shown below:*

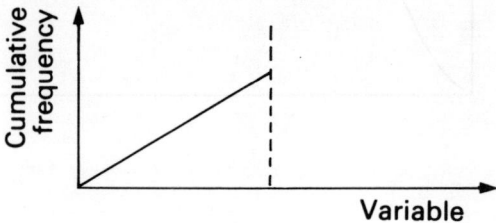

45. (a)

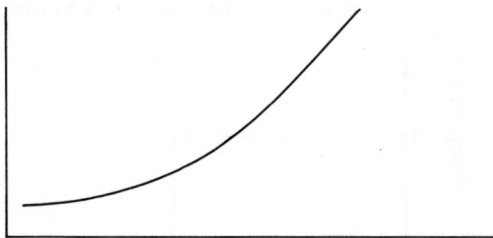

(b)

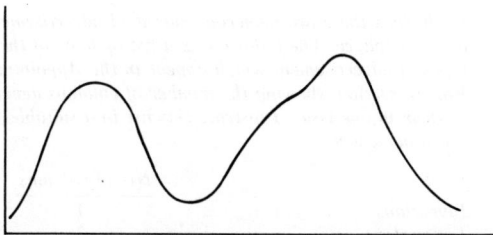

46.

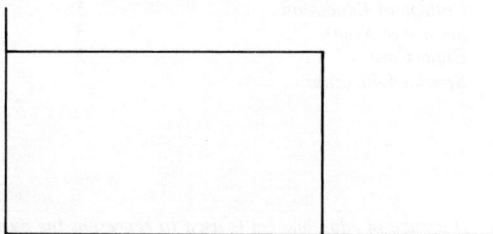

47. *A continuous variable, x, has the frequency curve shown below. Sketch the corresponding cumulative frequency curve.*

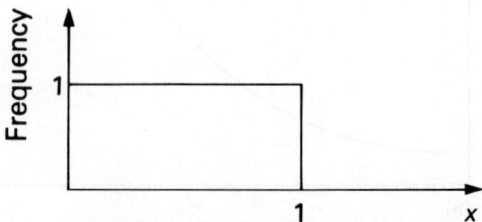

48. *Each week there are numerous classified advertisements in a newspaper. The following is a list of some of the types of advertisement which appear in the Appointments Vacant section, showing the number of columns devoted to them in one issue. Illustrate this list in a suitable way, e.g. a bar chart.*

Type	No. of columns
Librarians	1
Fellowships	2
Universities	10
College of Education	8
Service of Youth	3
Child Care	2
Special Education	7

49. *If a cube of edge one cm is used to represent the output of 1,000 tons of steel, what would be the output represented by a cube of edge three cm?*

47.

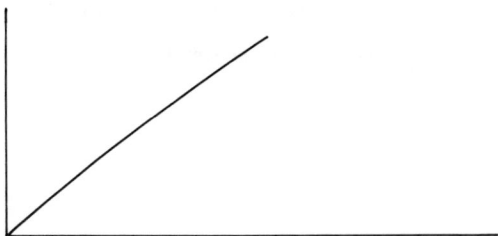

48.

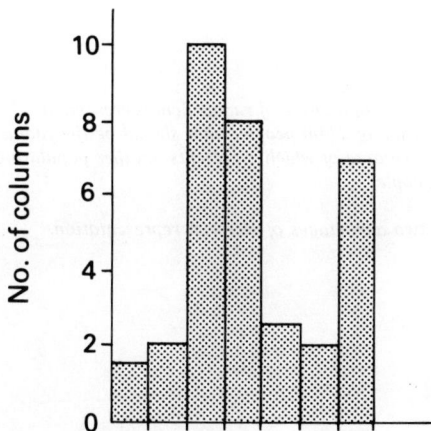

49. 27,000 tons (since $3^3 = 27$)

50.

Age (years)	Less than 5	5–	10–	20–	40–70
Number	15	16	20	10	6

Sketch a histogram to represent this data.

51. If the area of a circle of radius 2cm represents a population of 2,000 people, what should be the radius of a circle the area of which represents another population of 500 people?

52. Give two advantages of pictorial representation.

50.

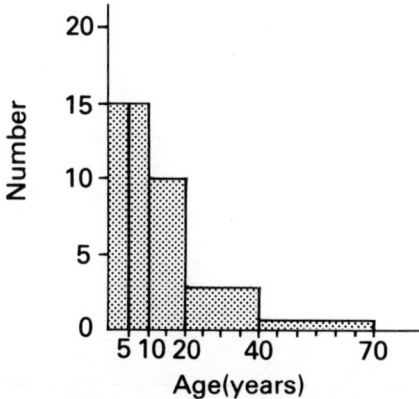

51. 1cm (Proof: Area of circle = $\pi r^2 = \pi 2^2 = 2,000$ people, therefore $4\pi = 2,000$ people therefore 500 people will be represented by $4\pi/4 = \pi r^2$, therefore $r^2 = 1$ and $r = 1$cm.)

52. (a) It is a clear and effective way of interpreting and presenting a subject visually.
    (b) It may help to clarify a problem or reveal facts which were not obvious in the original data.

53. Sketch the outline of a negatively skew histogram.

54.

Salary	less than £2000	£2000–	£2400–	£3000–	£4000–	£6000–£10000
No. of earners	7	22	51	69	45	17

The above frequency table relates to the salaries of a particular group of civil servants. If, when plotting the data as a histogram, the height of the bar representing the salary group £2000– is 22 units, what should be the height of the bar representing the salary group £4000–?

55. Explain what is meant by a negatively skewed distribution.

53.

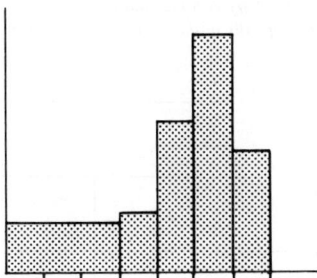

54. 9 units (for a width of £400, one unit is used per number; for a width of £2,000, 1/5 unit is used per number).

55. A distribution in which the mode has the highest value and the mean the least value.

Questions 56–58 concern the following histogram, which represents a stock of nails held by a handyman.

Column X shows a total of 200 nails for the class 0–5cm.

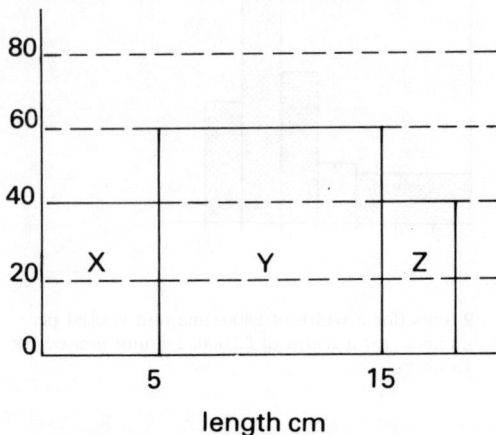

length cm

56. How many nails are represented by Column Y?

57. If Column Z represents 120 nails, what is the upper boundary (end point) of the class 15cm upwards?

58. What is the total number of nails in the handyman's stock?

56. 600 nails ($5 \times 40 = 200$ therefore $10 \times 60 = 600$)

57. 18cm ($x \times 40 = 120$ therefore $x = 3$ therefore upper boundary $= 15 + 3 = 18$)

58. 920 ($200 + 600 + 120$)

*Questions 59 and 60 concern the following statistics, which
show the age groupings of people in a small village.*

Age (years)	Frequency
Less than 5	10
5–17	45
18–64	168
65 and over	47

59.   *If a Pie Chart were drawn to represent this information,
      what would be the angle of the sector representing
      "18–64 years"?*

60.   *Which age-group would be represented on the Pie Chart
      by an angle of 60°?*

61.   *State the main purposes of representing numerical data in
      diagrammatic form.*

59. 224° (total frequency of 270 = 360° ∴ 1 unit = $\frac{360}{270} = \frac{4}{3}$
∴ 160 = $\frac{168}{1} \times \frac{4}{3}$ = 224)

60. 5–17 (60° = $\frac{1}{6}$ of 360° = $\frac{1}{6}$ of 270 = 45 = age group 5–17)

61. (a) presents it more attractively
    (b) makes data easier to understand at a glance
    (c) allows comparisons to be made easily.

62. Sketch each of the following statistical diagrams:
     (a)   A Pie Chart
     (b)   A Sectional Bar Chart
     (c)   An Array

63. What is the main characteristic of a histogram?

62.

(a)

(b)

(c)

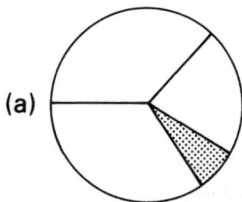

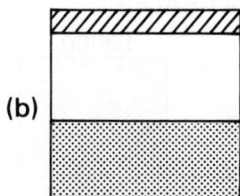

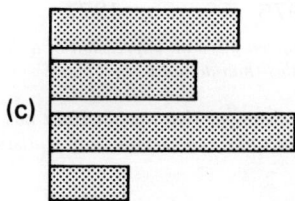

63. The areas of the rectangles must be proportional to the frequency in the classes represented by the rectangles.

64. The value of sales per year of a dog-food called Woofmeat were as follows:

Year	Sales (£)
1975	1,000
1976	2,000
1977	3,000

In an advertising campaign, the makers of Woofmeat produced the following diagram:

WOOFMEAT SALES

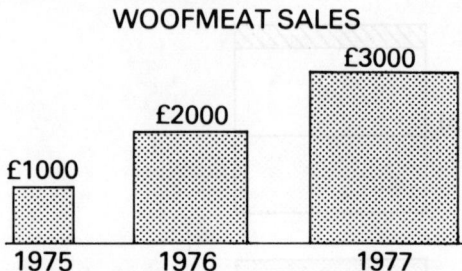

Explain how the diagram gives an impression of a much larger increase in sales than actually took place.

65. Explain briefly one advantage that the bar chart has over the pie chart in the pictorial representation of statistical data.

66. What are ratio-scale graphs?

64. The ratio of the sales is $1:2:3$ but the squares drawn have areas in the ratio of $1^2:2^2:3^2$ i.e. $1:4:9$.

65. It is possible to obtain a clearer idea of each component in a bar chart than in a pie chart since it is often difficult for the eye to distinguish between angles in a pie chart.

66. These are graphs which have the vertical scale proportional not to the numbers themselves (as with ordinary graphs) but to the logarithms of the numbers. These graphs show relative, not absolute, changes.

67. *Summarise the features of the ratio-scale graph.*

68. *Determine the total number of persons whose income distribution is shown in the histogram below.*

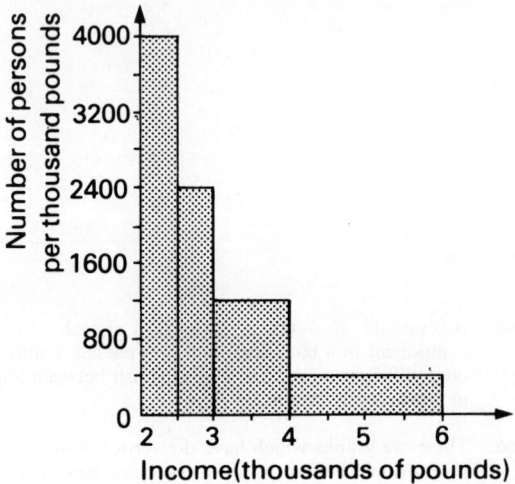

67. (a) The horizontal axis is normal.
    (b) The vertical axis has a logarithmic scale and a number shown on this axis will have its logarithm plotted.
    (c) Equal vertical distances represent <u>equal proportional</u> rates of change and hence the slope of the line connecting plotted points indicates the rate of change.

68. 10,400 [4,000 + 2,400 + (2 × 1,200) + (4 × 400)]

69. *Explain briefly one advantage that the pie chart has over the bar chart in the pictorial representation of data.*

70. *Point out any deficiencies in the following diagram.*

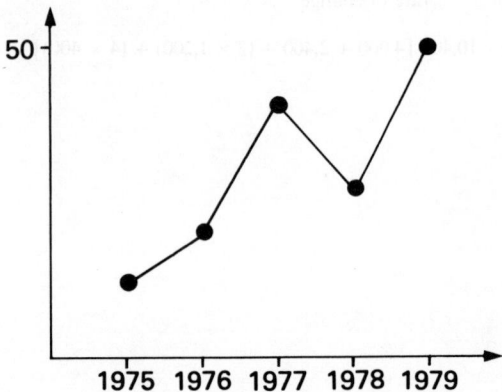

71. *A pie chart, showing exports for a given year, has a sector of 21° representing exports valued at £240 million. What is the value, to two significant figures, of all exports?*

69. It is possible to see the proportion of any given item of the total items at a glance.

70. (a) Neither axis is labelled.
    (b) The vertical axis has only one point marked and does not indicate whether or not the value at the base of that axis is zero.
    (c) There is no title.
    (d) The units on the vertical axis are not stated.

71. 4,100m $(21° = £240m$ $\therefore$ $1° = \frac{240}{21}$
    $\therefore$ $360° = \frac{240}{21} \times \frac{360}{1} = 4,114 \simeq 4,100)$.

72. *Give an example of the type of data which may be best illustrated by a sectional bar chart as shown below:*

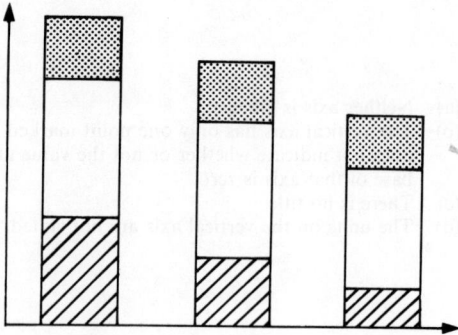

72. Numbers of different types of shops in a town each year.

73. On the diagram, complete the histogram corresponding to the data given below:

Income	£2,000–	£2,500–	£3,000–	£4,000–£6,000
No. of persons	2000	1200	1200	800

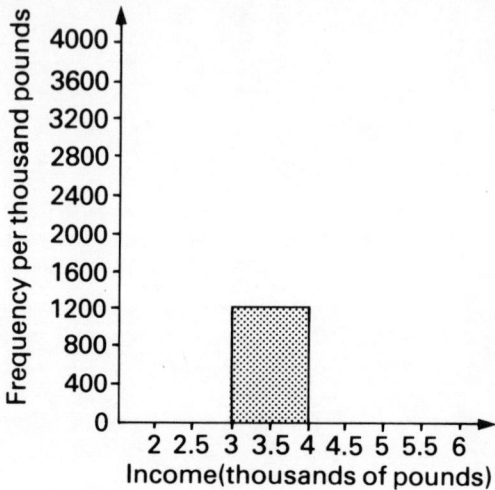

73.

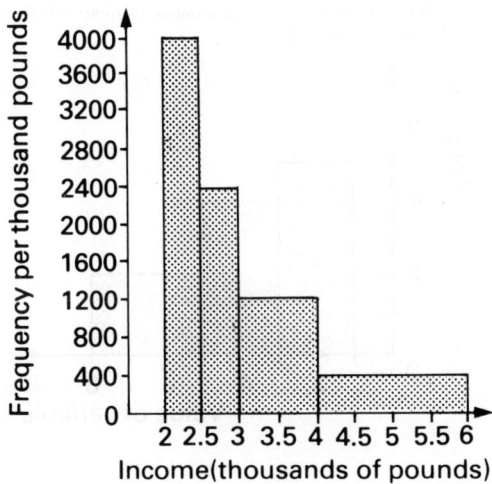

74. *Given that the histogram indicates that 32,000 items have a value less than 3, find the number of items which have a value greater than 2.*

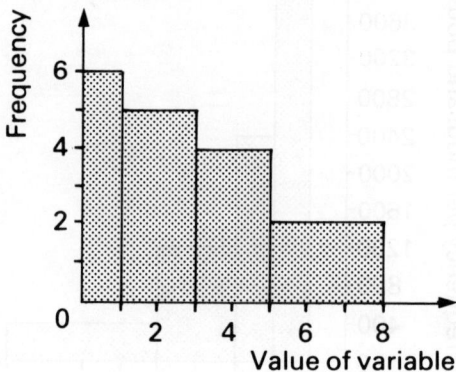

74. 38,000 [$(1 \times 6) + (2 \times 5) = 16$ units which represent 32,000; $(1 \times 5) + (2 \times 4) + (3 \times 2) = 19$ units. 1 unit represents 2,000; 19 units represent 38,000].

75. *The dimensions of the bottle representing the milk production of county A are three times those of the bottle representing the milk production of county B. What is the ratio of the volume of milk produced by county A to that produced by county B?*

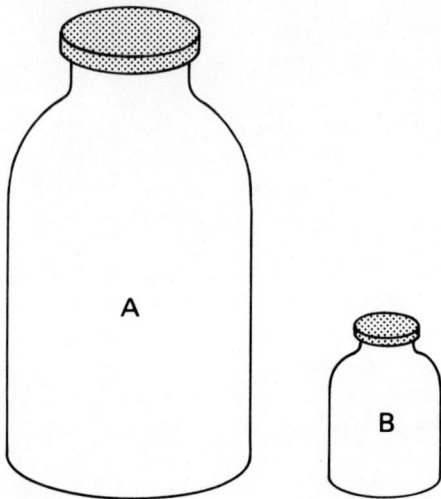

A

B

75. The illustrations represent volume hence if the dimensions are in the ratio of 3:1, their volumes are in the ratio of $3^3:1^3$, hence the ratio of the volume of milk produced by A to that produced by B is 27:1.

76. No. per year

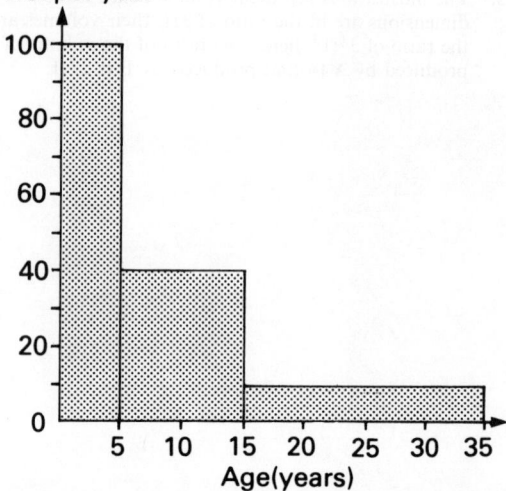

Complete the table given below to correspond with the histogram shown.

Age(years)	0-	5-	15-35
Frequency			200

(As 20 × 10 = 200; 10 × 40 = 400 and 5 × 100 = 500.)

76.

Age (years)	0–	5–	15–35
Frequency	500	400	200

77.

Height (cm)	150–	160–	180–
Frequency	40		

Part of a distribution table is shown above. On the histogram representing this table the height of the Column corresponding to "150–" was 10cm. If the height of the Column corresponding to "160–" was 8cm calculate the frequency.

78. *Pie charts are to be drawn to illustrate and compare the acreage under cereal crops in 1960 and 1970. What will be the radius of the 1970 chart if the radius of the 1960 chart is 10cm?*

	Acres (100s)	
Cereal	1960	1970
Wheat	20	36
Barley	22	28
Oats	33	44

77. 64 (Column "150–" has width 10 and height 10 ∴ 100 represents a frequency of 40; column "160–" has width 20 and height 8 ∴ 160 represents a frequency of $\frac{40}{100} \times \frac{160}{1} = 64$)

78. 12cm (In 1960 total of 75 represented by area of 100π; ∴ total of 1 represented by area $= \frac{100}{75}$; ∴ total of 108 for 1970 represented by area $= \frac{100}{75} \times \frac{108}{1}\pi$ $= 144\pi$. Square root of 144 = 12).

79. The pie chart shown represents the sales, in £, of various types of meat, made by a certain butcher during one week.

The angle of the sector representing sales of poultry is 20°. The butcher sells lamb to the value of £30 but sells poultry only to half this value.

Estimate, for the week:
(a) his sales of pork,
(b) his total sales.

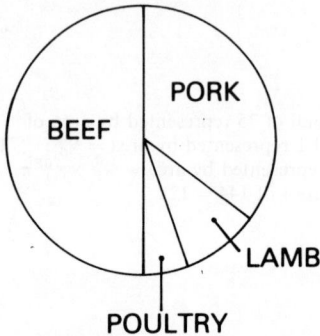

80. A small firm decides to manufacture shoes. It has only sufficient capital to install machinery to manufacture one size. In deciding which size to manufacture, is the firm likely to be most interested in the arithmetic mean, the median or the mode of sizes of shoes worn by the population? Give a reason for your answer.

79. (a) £90 (Since 20° represents £15; 40° represents £30 and pork is $180° - (20° + 40°) = 120°$)
    (b) £270 (Since poultry is £15; lamb is £30; pork is £90 and this total of £135 represents half his total sales)

80. The mode since this would represent the size worn by more people than any other size.

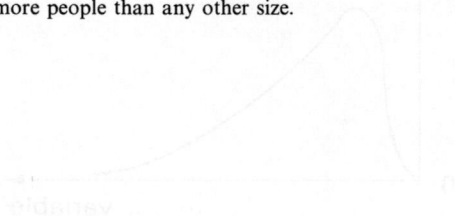

81. The mean of the following set of numbers is 5 and the mode is 4. What values have x and y?

$$3 \quad 7 \quad x \quad 9 \quad 4 \quad 2 \quad y$$

What is the median?

82. A small company is about to start production of an item of clothing. The company is initially limited to producing items of one particular size (owing to the cost of installing machinery, etc). If you were in charge of production which population average size—the mean, the median or the mode—would you be most interested in estimating?

83. Name a type of distribution for which the mean, the median and the mode have the same numerical value.

84. Which of the three common measures of location, the mean, the median and the mode, may be badly affected by an abnormal individual item?

85. Given the type of distribution illustrated below, which of the three common measures of location—the mean, median and mode—has the smallest value?

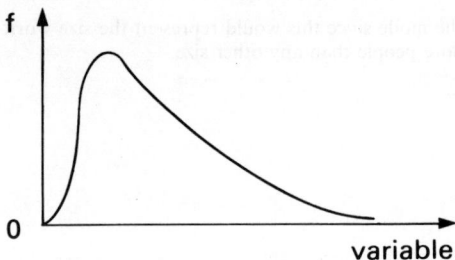

81. $x = 4$ or 6; $y = 6$ or 4 (since the mode is 4 either x or y must be 4. The other value must be $(7 \times 5) - 29 = 6$) median $= 4$ (2, 3, 4, ④, 6, 7, 9).

82. The mode since this indicates the most popular size.

83. A completely symmetrical distribution.

84. The mean.

85. The mode.

86. *The arithmetic mean height of a class of 20 boys is 154cm (to the nearest cm). Another boy, whose height is 168cm, joins the class. What is the arithmetic mean height of the class now?*

87. *What is the arithmetic mean of the integers 1, 2, 3, . . . , 9?*

88. *If a man walks 1 kilometre at 3km/h and 2 kilometres at 4km/h, What is his average speed for the three kilometres?*

89. *A cricketer's batting average in each of the last two seasons was 48.0 and 37.6 runs respectively. If the corresponding numbers of completed innings were 25 and 30, what was his batting average over two seasons?*

90. *What is the arithmetic mean of 1 8 3 4 3 5 8 10 3?*

91.

No of Houses	Weekly Rental
45	£10
35	£8
20	£5

The above figures show the weekly rent of houses on an estate. Calculate the mean weekly rent per house.

92. *Fifty students sat an examination. Their marks are tabulated below.*

Mark	1–10	11–20	21–30	31–40	41–50
No. of Students	4	11	15	16	4

Estimate the arithmetic mean.

86. 155cm (i.e. $(20 \times 154) + 168$ all divided by 21)

87. 5 (i.e. $1 + 2 + 3 + \ldots 9$ all divided by 9).

88. 3.6km/h (i.e. $3 \div (\frac{1}{3} + \frac{1}{2})$)

89. 42.327 (i.e. $(48 \times 25) + (37.6 \times 30) \div 55$).

90. 5 (i.e. $1 + 8 + 3 + \ldots \div 9$).

91. £8.3 (i.e. $(45 \times 10) + (35 \times 8) + (20 \times 5) \div 100$).

92. $26\frac{1}{2}$ (Find mid points; assume a mean, calculate deviations of mid points from assumed mean and multiply by frequency. Divide by 50 and add to the assumed mean.)

93. *What advantages does the arithmetic mean have?*

94. *Outline two disadvantages of the arithmetic mean.*

95. *The mean of one set of six numbers is $8\frac{1}{2}$ and the mean of a second set of eight numbers is $7\frac{1}{4}$. Calculate the mean of the combined set of 14 numbers.*

96. *In 14 consecutive innings during a season a batsman's scores were*

    17 48 21 30 94 98 0 5 120 73 60 184 25 13

    Find his average score.
    Before the end of the season the batsman had 8 more innings and thereby decreased his average for the season by 4.2 runs. Find his average score for the last 8 innings.

97. *The table below shows the number of bottles of milk delivered to each of 200 families in a particular day. Calculate the mean.*

No. of bottles delivered	1	2	3	4	5	6
No. of families	54	72	37	20	11	6

98. *A man runs 1500 metres in 250 seconds. If both these quantities are accurate to 1% between what limits does his average speed lie?*

99. *A motorist travels for 30 minutes at an average speed of 40km/h then for a further 20 minutes at 60km/h. What must be his average speed for the next 40 minutes if he wishes to attain an average speed of 50km/h for the whole journey?*

93. (a) It is easy to understand and calculate, and is commonly used.
    (b) It makes use of all the data in the group and can be determined, therefore, with mathematical exactness.
    (c) It can be determined when nothing more is known than the total value or quantity of the items, and the number of them are known.

94. (a) It may give undue weight to, and be unduly influenced by, extreme abnormal items.
    (b) The average may be a value which does not correspond with a single item.

95. $7\frac{11}{14}$ (i.e. $(6 \times 8\frac{1}{2}) + (8 \times 7\frac{1}{4}) \div 14$)

96. Average score is 56.3 (i.e. $17 + 48 + \ldots \div 14$)
    Average for last 8 innings is 44.75 (i.e. 52.1×22 minus previous total, divided by 8).

97. 2.4 (i.e. $(1 \times 54) + (2 \times 72)$ etc $\div 200$)

98. Between 5.88 metres per second and 6.12 metres per second $\left(\text{i.e. between } \dfrac{1485}{252.5} \text{ and } \dfrac{1515}{247.5} \right)$

99. $52\frac{1}{2}$km/h (Since he does 40km in 50 minutes and he needs to do 75km in 90 minutes; hence in 40 minutes he must do 35km, therefore in 60 minutes he must do $52\frac{1}{2}$km.)

100. *State briefly the assumption that is made when the mean is calculated from a grouped frequency table.*

101. *The table below gives the distribution of scores obtained when a die was thrown 40 times.*

Score	1	2	3	4	5	6
Frequency	5	8	7	6	4	10

If a series of fives is then thrown, how many would be required for the overall arithmetic mean to be 3.8?

102

Height (cm)	152	153	154	155	156	157
Frequency	4	7	9	5	4	1

The frequency distribution given above gives the height, correct to the nearest cm, of 30 boys. Calculate the modal height.

103. *What is the mode of the following set of numbers?*

4   7   11   14   16   16   31

104. *Define the mode*

105. *What is the mode of 1 8 3 4 3 5 8 10 3?*

100. The assumption is that the mean is not too distant from the median.

101. 5

102. 154cm (i.e. the height representing the greatest frequency).

103. 16 (it occurs twice; other figures occur only once).

104. The term used to designate the most frequent item in a series; the value which occurs most in a group; the position of greatest density.

105. 3 (it occurs three times whereas 8 occurs twice and other figures only once)

106. *Estimate the mode of the distribution illustrated below.*

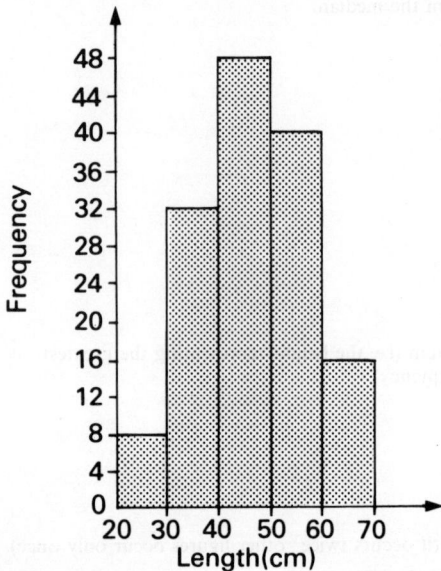

107. *What advantages does the mode possess as a form of average?*
108. *Outline the disadvantages of the mode.*
109. *The results of throwing a die 100 times are shown below. Determine the mode of this distribution.*

Number thrown	1	2	3	4	5	6
Frequency	16	20	13	18	18	15

106. 48 (By drawing a free-hand smoothed curve approximately midway between points a smoothed histogram is formed, the highest ordinate of which indicates the mode.)

107. (a) It is easy to understand.
(b) Extreme items do not affect its value.
(c) Only the middle items need be known.

108. (a) It is often not clearly defined.
(b) Exact location is often uncertain.
(c) Arrangement of data as an array, and the regrouping of items (when required) is tedious.
(d) It cannot be used for arithmetic calculations (e.g. to determine the total number or value of all the items).

109. 2 (Since it occurs more frequently than any other value.)

110. *Find the modal class of the following distribution.*

Length (cm)	22–	26–	28–	29–	30–31
Frequency	12	12	9	11	6

111. *Describe a situation in which the mode is a more useful average than either the mean or the median.*

112. *What is the median of the following set of numbers?*

$$28 \quad 13 \quad 35 \quad 22 \quad 7 \quad 12 \quad 24 \quad 31 \quad 27$$

113.

No. of days	0–	50–	100–	150–	200–	250–	300–365	Total
No. of children	29	24	30	20	25	17	11	156

The above table gives the ages in days (over 15 years old) of children in the fifth form at a particular school. What is the median age of children in the fifth form at that school?

114. *What is the median?*

115. *What is the median of 1 8 3 4 3 5 8 10 3?*

110. Classes 22–26 and 26–28 are both modal classes.

111. A manufacturer of shoes will find the arithmetic mean useless since it is unlikely to represent any size; the median represents only the middle of a range whereas the mode will represent the most popular size.

112. 24 (Since this is the middle number when they are placed in order.)

113. 15 years 115 days.

114. The median of a set of numbers arranged in an array is the middle value, or the mean of the two middle values. It is the central value dividing an array.

115. 4 (The middle number when the numbers are arranged in order.)

116. *Fifty students sat an examination. Their marks are tabulated below.*

Mark	1–10	11–20	21–30	31–40	41–50
No. of students	4	11	15	16	4

Estimate the median mark.

117. *What are the advantages of the median as an average?*

118. *Outline the disadvantages of the median.*

119. *The table below gives the family size of 120 families in a particular area. Determine the median.*

No. of persons in family	1	2	3	4	5	6	7 and over
No. of families	18	29	26	21	10	9	7

120. *Twenty observations of a variable size which takes integral values are placed in order. If the 8th observation is 17 and the median value is 17, what can be said about the observations?*

116. 27 2/3rds.

117. (a) If found directly, it is representative of an actual item.
    (b) It is simple to understand.
    (c) Extreme items do not effect its value.
    (d) It can be obtained even when the values of all items are not known. Provided the middle items are known, and that there are the same number of larger and the same number of smaller items, the median can be located.
    (e) It can be used for measuring qualities and factors to which mathematical measurement cannot be given.

118. (a) If the items are few, it is not likely to be representative.
    (b) If the distribution is irregular, the location of the median may be indefinite.
    (c) When the items are grouped (as into classes), it cannot be located with precision, particularly in a discrete series.
    (d) The arranging of data in the necessary array is often tedious.
    (c) It cannot be used to determine the total value of all the cases or items. The number of items multiplied by the median will not give the total of the data. It is not suitable for arithmetic calculations, and has limited value in practical work.

119. 3 (The median family is the $60\frac{1}{2}$th family which has 3 persons.)

120. If the 8th observation is 17 and the median is 17, then the 9th, 10th and 11th observations are all 17.

121. *How is the geometric mean calculated?*

122. *What advantages does the geometric mean possess?*

123 *Outline the disadvantages of the geometric mean.*

124 *Give one advantage that the geometric mean has over the arithmetic mean as a measure of location.*

125. *Give three measures of dispersion.*

121. The geometric mean is the nth root of the product of n items in a group e.g. the geometric mean of 10; 100 and 1000 is

$$\sqrt[3]{10 \times 100 \times 1000} = \sqrt[3]{1,000,000} = 100$$

122. (a) It makes use of all the data in a group, and can be calculated with mathematical exactness, provided that all the quantities are greater than zero and positive.

(b) It is the only average that can be used to indicate rate of change e.g. from 1977 to 1979 prices increased by 5%, 10% and 15.7% respectively. The average annual increase is the geometric mean i.e.

$$\sqrt[3]{5 \times 10 \times 15.7} = \sqrt[3]{785} = 9.2\%$$

(c) Large items have less effect on it than on the arithmetic mean.

123. (a) It is impossible to use it when any items are zero or negative.

(b) It is more difficult to calculate and less easily understood.

(c) It may locate a value which does not correspond with any actual example.

124. It is less affected by large items.

125. Any three from:
(a) range
(b) quartile deviation
(c) standard deviation
(d) mean deviation

126. *Forty packets of sugar are taken from a production line. It is found that their weights differ slightly. The heaviest packet is replaced by an even heavier packet. Name a measure of dispersion which would be exactly the same before and after the experiment.*

127. *Name the measure of dispersion which is the average difference of each of a set of observations from their mean value.*

128. *What is the range of the numbers?*

$$1 \quad 8 \quad 3 \quad 4 \quad 3 \quad 5 \quad 8 \quad 10 \quad 3?$$

129. *The blood pressures, measured to the nearest mm, of 200 men are tabulated below:*

Blood pressure	70–74	75–79	80–84	85–89
No. of men	28	36	103	33

Find the maximum possible value of the range of this distribution.

130.

<u>Age Groupings of People in a Small Village</u>

Age (years)	Frequency
Less than 5	10
5–17	45
18–64	168
65 and over	47

In which age-group is the upper quartile?

126. Quartile deviation (since it is concerned only with the value of the upper quartile and of the lower quartile).

127. Mean deviation.

128. 9 (i.e. the difference between the highest and lowest numbers)

129. 19.9 (i.e. the difference between 89.4 and 69.5)

130. 18–64 (the cumulative frequency is 270 at the end of this group; the upper quartile is the $\frac{3(271)}{4}$th item).

131. The following table shows the milk yield, in litres, of 100 cows in a year. Estimate the upper and lower quartiles and the semi-interquartile range.

Milk yield	1500–	2000–	2500–	3000–	3500–	4000–5000
No. of cows	14	25	34	20	5	2

132. The diagram below shows the cumulative frequency curve of the examination marks of 400 candidates. Estimate the interquartile range.

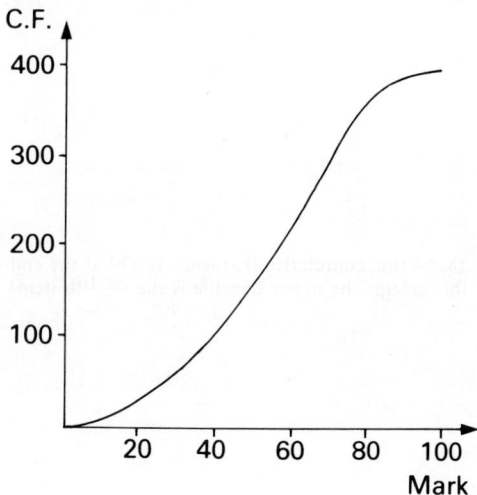

131. 3247 and 2225; 511

132. Approx 27 (upper quartile is value of 300th item i.e. 67;
      lower quartile is value of 100th item i.e. 40; difference
      = 67 − 40 = 27).

133. *Fing the mean deviation of the following set of observations:*

$$5 \quad 8 \quad 11 \quad 2 \quad 4 \quad 6$$

134. *Calculate the mean deviation from the <u>median</u> of the following distribution:*

No. of flowers per plant	8	9	10	12	14	17
No. of plants	3	9	4	2	1	1

135. *The mean of the three numbers 2, 5 and x is greater than 5. The mean deviation is 4. Calculate the value of x.*

136. *Explain why the mean deviation is used less frequently than the standard deviation as a measure of dispersion.*

137. *A group of 10 observations has an arithmetic mean 13.2 and mean deviation from the mean 1.5. Two further observations are such that the arithmetic mean and mean deviation from the mean of all 12 observations is the same as before. What are the values of these two observations?*

138. *It is given that the numbers*

$$-4 \quad -2 \quad -1 \quad 0 \quad a \quad b \quad c$$

are in ascending order and that their mean is zero.
  (i) *What is the sum of the numbers a, b and c?*
  (ii) *Calculate the mean deviation from the median of the seven numbers.*

133. 2 1/3rd (mean = 6; $\Sigma d = 14$; n = 6)

134. $1\frac{3}{10}$ (median = 9; $\Sigma f = 20$; $\Sigma fd = 26$)

135. 13

136. The mean deviation cannot be used in further statistical analysis whereas the standard deviation is of the greatest importance in later statistical work.

137. 14.7 and 11.7 (They must total 26.4 to maintain the same mean; their differences from the mean must total 3.)

138. (i) 7   (ii) 2

139. The standard deviation of a set of 5 numbers is 7.0 and the standard deviation of another set of 15 numbers is 6.0. Given that the two sets of numbers have the same mean, calculate the standard deviation of the combined set of 20 numbers.

140. Given that the standard deviation of the integens from 1 to 10 inclusive is 2.87, find the standard deviation of the integens 11–20 inclusive.

141. Using the information in Q. 140, find the standard deviation of the first ten even integens.

142. Give an advantage that the standard deviation has over the interquartile range as a measure of dispersion.

143. The arithmetic mean of the following distribution is 10 flowers per plant. Calculate the standard deviation.

No. of flowers per plant	8	9	10	12	14	17
No. of plants	3	9	4	2	1	1

144. The arithmetic mean and standard deviation of a set of examination marks are 25 and 10 respectively. If the marks are adjusted so that the new arithmetic mean becomes 60 and the standard deviation 20, find the adjusted mark corresponding to an original mark of 30.

145. What is the variance of the integens 1, 2, 3, ..., 9?

146. Give one advantage that the standard deviation has over the range as a measure of variability.

139. $6.3 \left(\text{Since } \sqrt{\dfrac{\Sigma \, d^2}{5}} = 7 \quad \text{i.e. } \dfrac{\Sigma \, d^2}{5} = 49 \right.$

$\text{i.e. } \Sigma \, d^2 = 245 \text{ and } \sqrt{\dfrac{\Sigma \, d^2}{15}} = 6 \quad \text{i.e. } \dfrac{\Sigma \, d^2}{15} = 36$

$\text{i.e. } \Sigma \, d^2 = 540 \quad \therefore \text{ combined } \Sigma \, d^2 = 795$

$\left. \text{and } \therefore \sigma = \sqrt{\dfrac{795}{20}} = 6.3 \right)$

140. 2.87

141. 5.74 (double the previous figure)

142. The standard deviation makes use of all the items.

143. $2.17 \left(\text{i.e. } \sqrt{\dfrac{\Sigma \, fd^2}{\Sigma f}} = \sqrt{\dfrac{94}{20}} \right)$

144. 70

145. $6\frac{2}{3}$ ($\bar{x} = 5$; $\Sigma \, d^2 = 60$; $n = 9$)

146. The standard deviation makes use of all the items; the range is concerned only with the two extreme items.

147. A group of students was given a test which was marked out of a possible total of 30. The standard deviation of the marks obtained was 5.4. What could be the standard deviation if the marks were scaled and expressed as percentages?

148. Give a formula for calculating the standard deviation.

149. What is the formula for the coefficient of variations?

150. What is "variance" and what is its usefulness?

151. Determine the variance of the following observations:

$$-3 \quad 0 \quad 1 \quad 1 \quad 6$$

152. Find the variance of five consecutive odd integens.

153. The following instructions show how to calculate the standard deviation but they are not in the right order. Rewrite the steps in the correct order.

Average the squares of the differences.
Take the square root.
Calculate the mean.
Square the differences.
Calculate the differences from the mean.

154. Year   1978   1979
    Index No.   105.0   107.1   (1976 = 100)

Calculate a new index number for 1979 with 1978 = 100.

155. The prices of a commodity in 1977 and 1979 were 80p and 89p respectively. What is the price relative for the commodity in 1979 if 1977 = 100?

147. 18

148. $\sqrt{\dfrac{\Sigma d^2}{n}}$ or $\sqrt{\dfrac{\Sigma fd^2}{\Sigma f}}$ or $\sqrt{\dfrac{\Sigma fd^2}{\Sigma f} - \left(\dfrac{\Sigma fd}{\Sigma f}\right)^2}$

149. $\dfrac{\sigma}{x} \times \dfrac{100}{1}$ i.e. $\dfrac{\text{standard deviation}}{\text{mean}} \times \dfrac{100}{1}$

150. The variance is the square of the standard deviation. Its usefulness lies in the fact that, unlike deviations, variances can be added. If two variances of equal-sized distributions are added together, the result would be the combined variance of the two distributions.

151. 8.4 ($\bar{x} = 1$; $\Sigma d^2 = 42$; $n = 5$)

152. 8 ($n = 5$; $\Sigma d^2 = 40$)

153. (i) Calculate the mean.
(ii) Calculate the differences from the mean.
(iii) Square the differences.
(iv) Average the squares of the differences.
(v) Take the square root.

154. 102

155. 111.25

156.

Commodity	A	B	C	D
Index	101	103	105	102
Weight	2	3	4	2

Calculate a suitable index for the combined commodities.

157. *In January 1974, the Index for Fuel and Light was 100. A firm's bill for fuel and light for the first quarter of 1974 was £275. In the first quarter of 1977, using approximately the same amount of fuel and light, the firm received a bill for £550. Estimate the Index for Fuel and Light in January 1977.*

158. *In July 1977 the Food Index was 192, based on January 1974 = 100. Approximately how much money would be needed in July 1977 to pay for a parcel of groceries which, in January 1974, would have cost £25?*

159. *In the Index of Retail Prices, the item "food" has a weight of 314, whilst the item "Alcoholic Drink" has a weight of 63, State briefly why the first weight is so much larger than the second.*

160. *What is an Index Number?*

156. $103\frac{2}{11}$ $\left(\text{i.e. } \dfrac{\Sigma\,xw}{\Sigma\,w} = \dfrac{1135}{11} \right)$

157. 200

158. £48

159. "Food" has a weight so much larger than "Alcoholic Drink" since it represents a much larger proportion of income spent on that item and many more retail prices are concerned with food than with alcoholic drink. Basically it represents a much more important item.

160. An Index Number is a special kind of average and bears a relationship to a percentage. It represents a method of showing at a glance the overall direction of changes in a variable over a period of time.

161. Given that the weighted index for all commodities is 103,
find the index for Commodity D

Commodity	A	B	C	D
Index	102	104	104	
Weight	3	1	2	2

162. A firm uses four materials A, B, C and D in a chemical
process. The following table shows the prices, in £ per
tonne, of the materials in 1977 and 1979, and the average
weekly expenditure in 1977. Calculate an index number
of average weekly expenditure for 1979 based on 1977,
assuming that the amounts of A, B, C and D used remain
the same.

	Price (£ per tonne)		1977 Average
	1977	1979	Weekly Expenditure
			(£)
A	5	6	40
B	20	26	5
C	16	18	8
D	12	15	2

163. A certain article cost £64 in December 1977. By
December 1978 its price had increased by £8. From
December 1978 to December 1979 the price increased by
20%. Calculate the December 1979 index number using
December 1977 as base.

161.   103

162.   120.35

163.   135

164. *The price of a certain commodity in each of the years 1975 to 1978 is shown below.*

Year	1975	1976	1977	1978
Price	£2.00	£2.40	£3.00	£3.30

Starting with 100 for 1975 calculate index numbers for 1976, 1977 and 1978 using the chain base method.

165. *The index numbers, calculated on the chain base method, for a particular commodity are shown below. Calculate the 1979 index, to the nearest whole number, using 1977 as base.*

1977	1978	1979
100	107	108

166. *Explain briefly what is meant by a "weighted mean".*

167. *What is meant by the crude death rate?*

168. *What is a "corrected" or "standardised" death rate?*

164.   1976 = 120; 1977 = 125; 1978 = 110 (each year uses
       the previous year as base).

165.   116 (If 1977 = 100, then 1978 = 107 and 1979 = 107
       + 8% of 107)

166.   A weighted mean involves finding the arithmetic mean
       of a set of figures after having given a particular
       "weight" (according to importance) to each figure. The
       formula is $\dfrac{\Sigma xw}{\Sigma w}$.

167.   The crude death rate is the number of deaths per
       thousand of the population.

168.   A "corrected" or "standardised" death rate is the
       weighted average of the age-specific death rates using
       the standard age distribution as weights.

169. *Give a reason for the drop in the total unemployment figures in the summer months.*

170. *Give three possible components in a time series.*

171. *Explain briefly what is meant by seasonal variation and give an example where it may arise.*

172. *Find the value of the second moving average in a five point moving average for the following series:*

$$21 \quad 23 \quad 24 \quad 27 \quad 28 \quad 30 \quad 35 \quad 34$$

Questions 173 and 174 concern the following statistics which show the number of houses sold by an agent during the first six months of 1977.

Jan	Feb	Mar	Apr	May	June
14	9	13	17	21	23

173. *What is the value of the second 3-monthly moving average?*

174. *If the fifth 3-monthly moving average is 21, how many houses were sold in July?*

169. During the summer months a good deal of casual, temporary seasonal work is available e.g. in holiday resorts, on farms, etc.

170. Any three from:
Secular Variations
Seasonal Variations
Cyclical Variations
Random Variations

171. Seasonal variations is the tendency for figures to vary according to the season of the year e.g. in any monthly figures of annual ice-cream sales one would expect a higher figure in the summer months.

172. 26.4 (i.e. 23 + 24 + 27 + 28 + 30 all divided by 5).

173. 13 (i.e. 9 + 13 + 17 all divided by 3).

174. 19 [i.e. $(3 \times 21) - (21 + 23)$].

175. *Give an example of statistics forming a time series which are likely to show a seasonal variation.*

176. *The figures given below are observations taken at equal consecutive intervals of time.*

    4.6 4.4 4.1 4.5 4.0 3.8 3.5 3.9 3.4 3.2 2.9 3.3

    Find (a) the most appropriate number of observations to use as a basis for moving averages and hence (b) how many values of this moving average there will be.

177. *Recordings of a variable are taken once every day for two weeks and 7-day moving averages are calculated. How many values of the moving average are there?*

178. *Explain briefly what is meant by "random variation" in a time series.*

179. *Draw a scatter diagram showing high negative correlation between two variables.*

175. Figures of sales of bottles of sherry in a supermarket (far greater sales in the last quarter of each year).

176. (a) four
     (b) nine

177. 8

178. "Random variation" in a time series indicates irregular fluctuations such as those due to strikes, abnormal weather, war, etc. They are the residual factors which show up in line graphs after seasonal and cyclical variations have been eliminated.

179.

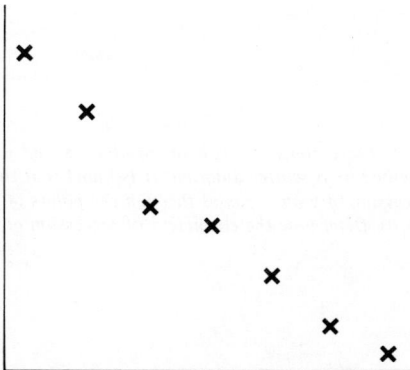

180.

x	3	4	7	7	9
y	5	3	4	6	7

The above table shows corresponding values of two variables x and y. If these points were plotted on a scatter diagram, state the co-ordinates of one point through which the line of best fit _must_ pass. (_Note:_ You are _not_ required to draw the diagram.)

181. Sketch a scatter diagram which demonstrates high (but not perfect) positive correlation between two variables.

182. A set of observations of a pair of variables, x and y, is represented by a scatter diagram. It is found that the line of regression of y on x passes through the points (4, 3) and (6, 0). Determine the coefficient of regression of y on x.

180. x = 6 and y = 5 (arithmetic means of x and y)

181.

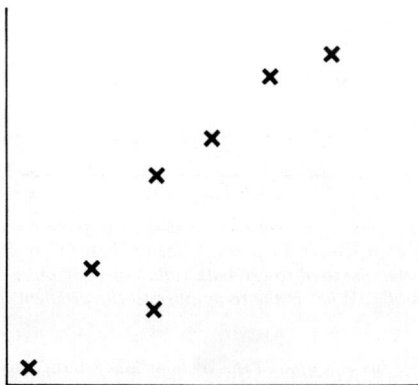

182. $-1\frac{1}{2}$

183. On the scatter diagram shown below mark the position of one further point in order that the coefficient of correlation between x and y is zero.

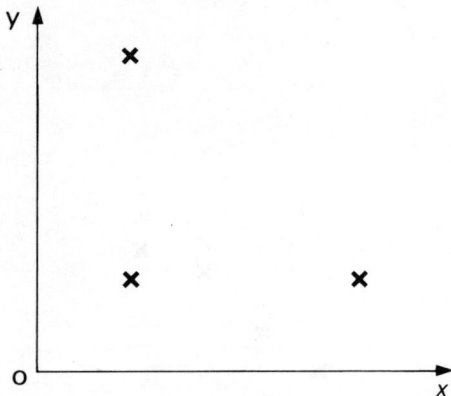

184. Observations of two variables x and y are plotted as points on a scatter diagram. A line of "best fit" is drawn which may be used to estimate values of y for given values of x. What is the term given to the gradient of this line?

185. What is the equation of the straight line relating the values of x and y given below?

x	1	2	3	4
y	3	5	7	9

186. What does a regression coefficient measure?

183.

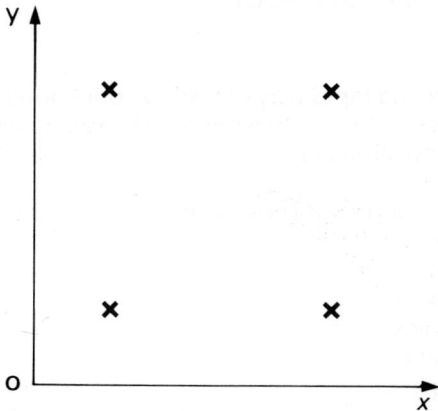

184. The coefficient of regression.

185. $y = 2x + c$ or $y = 2x + 1$

186. The regression coefficient measures the slope or gradient of the regression line.

TEST YOURSELF

An easy and rapid way of testing your knowledge of the basic facts. Questions and answers on the following subjects:

English Language (two books)
French
German
Commerce
Economics
Chemistry
Physics
Biology
Human Biology
Mathematics
Modern Mathematics
St. Matthew
St. Mark
St. Luke
St. John
Acts of the Apostles
Commercial Mathematics
Accounts
Statistics
British Isles Geography
British Economic History